CONTENTS

WHAT is SOLAR POWER?

Sunshine and solar power

Every fifteen minutes the Earth receives enough energy from the Sun to power everything on our planet for a whole year. Yet only a tiny amount of the energy we use in our daily lives comes directly from the Sun. At present, making electricity from sunlight is expensive.

Fossil fuels or solar power?

Oil, coal and gas are called fossil fuels. They give up their stored energy easily, when they are burnt. They can be used to make electricity cheaply.

Problems with fossil fuels

There are problems with fossil fuels:
- When they are burnt they pollute the air.
- We are going to use them all up. There will be none left.

Nuclear power, fossil fuels or solar power?

Nuclear power is cleaner than fossil fuels. However, it produces dangerous radioactive waste. We need to find cheaper ways to use solar power (the energy from sunlight). Sunlight is a clean and renewable source of energy. This book looks at ways in which it can be used.

This US Navy test platform gets electricity from solar panels (below) and a wind turbine (above).▼

SOLAR

's to be returned before
d

POWER

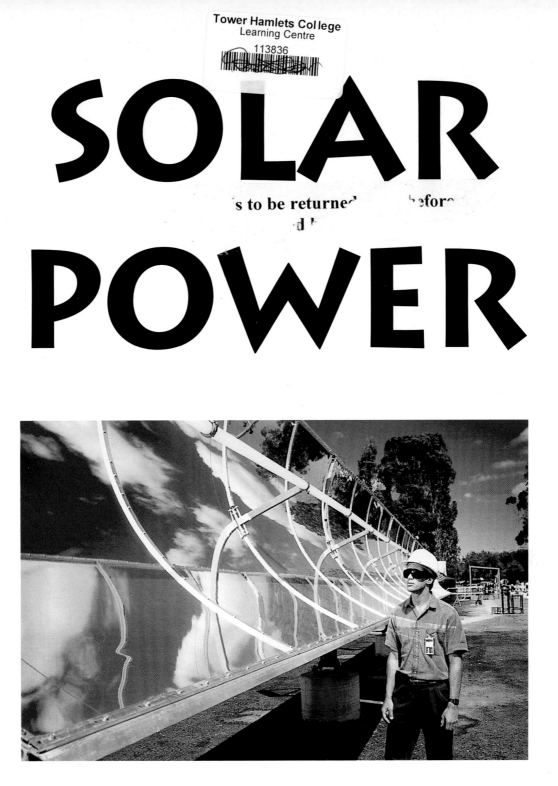

FIONA REYNOLDSON

WAYLAND

6.99

LOOKING AT ENERGY

OTHER TITLES IN THE SERIES

Fossil Fuels · Geothermal and Bio-energy
Nuclear Power · Water Power · Wind Power

This book is a simplified version of the title 'Solar Power' in Wayland's 'Energy Forever?' series.

Language level consultant: Norah Granger
Subject consultant: William Reynoldson
Editor: Belinda Hollyer   Designer: Jane Hawkins

Text copyright ©2001 Wayland
Volume copyright ©2001 Wayland

First published in 2001 by Hodder Wayland
an imprint of Hodder Children's books.

This paperback edition published in 2005
Reprinted in 2006 by Wayland, an imprint of Hachette Children's Books

British Library Cataloguing in Publication Data
Reynoldson, Fiona
Solar Power - (Looking at energy)
1.Solar energy - Juvenile literature
I.Title
333.7'923
ISBN-10: 0 7502 4723 1
ISBN-13: 978 0 7502 4723 8

Printed in China

Hachette Children's Books
338 Euston Road, London NW1 3BH

Picture Acknowledgments
Cover: main picture Klein/Hubert/Still Pictures, solar car Ecoscene/Anthony Cooper. Page 5 (Mitch Kezar). US Department of Energy: title page, pages 4, 7, 22-23, 22, 32-33, 34, 34 (bottom), 41. Ecoscene: pages 10, 25 (Brown), 30 (Erik Schaffer), 35 (John Farmer). James Davis Travel Photography: pages 9, 20-21. Frank Lane Picture Agency: page 13, Eye Ubiquitous: pages 12 (Brian Harding), 18 (Mark Newham), 44 (NASA). Mary Evans Picture Library: page 14. Stockmarket/Zefa: pages 15, 26, 36, 40-41 (Joe Sohm). UKAEA/AEA Technology: pages 19, 24, 31, 36-37. Ole Steen Hansen: 28, 29 (top). Shell Photo Library/Solavolt International: 29 (bottom). James Hawkins/Oxfam UK: PAGES 38, 38-39, 39.

Life on Earth
is only possible
because of energy
from the Sun. ▼

FACTFILE

FOSSIL FUELS
How much is left?

- about 50 years of oil.
- about 65 years of
natural gas.
- about 300 years of
coal.

THE SUN AS THE SOURCE

The Sun is so big that more than one million Earths would fit inside it. Because it is so big there are huge pressures inside it. These huge pressures make very high temperatures.

The Sun – a nuclear reactor

The Sun is like a nuclear reactor out in space. It is a nuclear reactor that works by nuclear fusion.

Violent eruptions from the Sun release huge bursts of energy and particles. Even a small eruption can be as powerful as a million nuclear bombs. ▼

WHAT IS NUCLEAR FUSION?
Nuclei hit each other so fast that they stick together. This releases energy and one neutron particle. (Nuclei is the plural of nucleus.) ▼

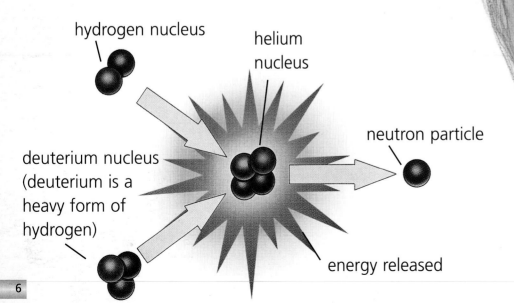

hydrogen nucleus

helium nucleus

neutron particle

deuterium nucleus (deuterium is a heavy form of hydrogen)

energy released

Nuclear fusion in the Sun

In the Sun, hydrogen atoms crash into each other. They crash so hard that they stick together. This makes a different element called helium. This type of reaction is called nuclear fusion (see the diagram on page 6). Huge amounts of energy turn into heat and light. This energy streams out into space. Some of this solar energy reaches the Earth.

(see the diagram on page 6)

FACTFILE

NUCLEAR FUSION

- Nuclear fusion can only happen when atoms move fast.

- Atoms only move fast at high temperatures.

- The temperature in the centre of the Sun is about 10 million degrees Celsius.

- We can make nuclear fusion take place in reactors, but only for one or two seconds.

Mirrors concentrate sunlight on to a transparent tube. The solar power helps to purify the contaminated water that is pumped through the tube. ▶

Electro-magnetic waves

Heat and light are made up of electrical waves and magnetic waves. These waves travel from the Sun like waves across the sea. Some of them hit the Earth. The light that leaves the Sun takes eight minutes to reach the Earth.

Most of the solar energy (heat and light from the Sun) is soaked up by the sea and land near the Equator. This land heats up so the water evaporates. This makes hot, sandy deserts.▼

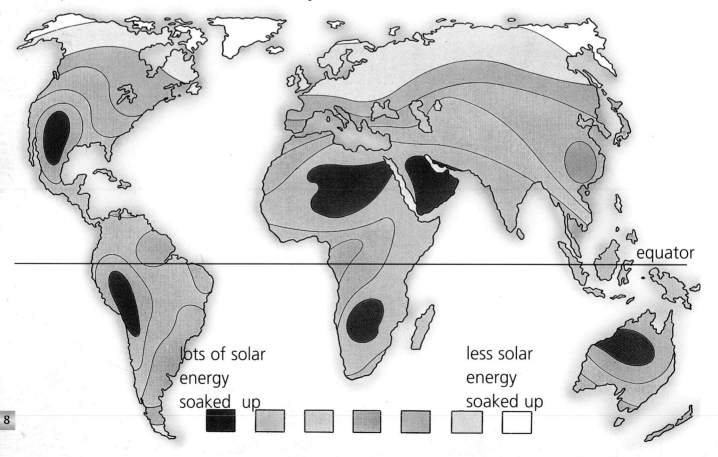

equator

lots of solar energy soaked up

less solar energy soaked up

FACTFILE

Light travels faster than anything else. It travels at a speed of 300,000 kilometres a second. This means we usually see things as soon as they happen. But light takes thousands of years to reach us from far away in the universe.

A rainbow happens when sunlight shines through rain. The raindrops split the light into seven colours: red, orange, yellow, green, blue, indigo and violet.

TOO MUCH SOLAR ENERGY

When the weather has been dry, trees dry out. Then they can catch fire in strong sunlight. In recent years, this has caused big forest fires in Spain, France and California in the USA.

STORED SOLAR ENERGY
Wood is stored solar energy. Burning wood releases this stored energy as heat. Coal is formed from trees that died millions of years ago.

10

These are trees in a rainforest in Australia. The green chlorophyll in their leaves captures the energy of the Sun, and uses it to make food for plants.

Plants and photosynthesis

All the plants and animals on the Earth rely on the Sun. There is green chlorophyll in the leaves of plants. This chlorophyll captures the Sun's energy. Plants use it to make food from water and carbon dioxide in the air. This process is called photosynthesis.

Solar wind

About half the Sun's energy turns into light. The rest of the energy is mostly infra-red heat. There are also small electrically charged particles that come from the Sun all the time. This is called solar wind. Storms around the Sun can make these particles blow out suddenly.

The solar wind particles

These particles shower the Earth. They can cause beautiful rainbow light in the sky near the North and South Poles. This coloured light is called an aurora.

Bad effects

However, the storms of solar wind particles can affect the Earth's magnetic field. This can cause problems to power cables, radio aerials, telephone lines and satellites.

SOLAR POWER in HISTORY

The Sun as a god

Many people in ancient times believed that the Sun was a god. For instance, the Ancient Egyptians believed that the Sun was a god who sailed across the sky in a heavenly ship.

In this photograph maize is drying in the Sun. People have always used the Sun to dry things, such as animals' skins, fish, leaves and coffee beans.

Eclipses

People in ancient times knew how important the Sun was in their lives. Solar eclipses are when the Sun disappears (partly or completely) behind the Moon. This was terrifying. Ancient peoples worried that the Sun might not come back.

Human sacrifices

Some people in ancient times killed animals or humans to please the Sun god. These were called sacrifices.

▲ This little metal carriage shows the Sun being pulled by a horse. It comes from Denmark and is at least 3,000 years old. It tells us that the people who lived in Denmark long ago probably worshipped the Sun as a god.

This is Mouchot and Pifre's solar-powered printing press. The dish-shaped mirror collected the solar power which heated a boiler. The boiler made steam to run the printing press.

FACTFILE

Archimedes lived more than 2,000 years ago. There is a story that he used a mirror to concentrate the Sun's rays. He aimed these rays at the enemy's ships. They caught fire. No one knows if he really did this. But it shows that he knew how to use solar power.

Using mirrors

In 1882 Augustin Mouchot and Abel Pifre invented a solar-powered printing press. A mirror dish concentrated the Sun's rays on to a boiler. This made steam. The steam operated an engine to power the printing press. It could print 500 copies of a newspaper every hour.

Uses of solar power in history

- In 1774 Joseph Priestley concentrated the Sun's rays to heat mercury oxide. It separated into mercury and oxygen. Oxygen was a newly discovered element.

- In the same century, Antoine Lavoisier built a solar-powered furnace. It was hot enough to melt metal.

- In 1891 Clarence M. Kemp invented a solar water heater. There are still half a million used today in California.

rays of light from the Sun

lens

hole — burning in paper

▲ On a sunny day, four square metres of ground soak up one kilowatt of solar power. This is enough to run an electric toaster.

A solar telescope that can use the Sun's energy to make measurements. ▼

▲ A lens can concentrate or focus the Sun's rays to burn a hole in paper.

SOLAR TECHNOLOGY

Simple solar collectors

There are several kinds of cheap, simple solar collectors.

- A cold frame is like a small greenhouse. The Sun's rays pass through the glass to heat the air inside so that plants grow quickly.

- A solar drier can dry vegetables (see the diagram below).

- A sheet of plastic can collect drinking water in the desert (see the opposite page).

A SOLAR DRIER
The Sun shines on the glass panel. Cool air flows under it and is warmed up. The warm air then flows through the box and dries the vegetables. ▶

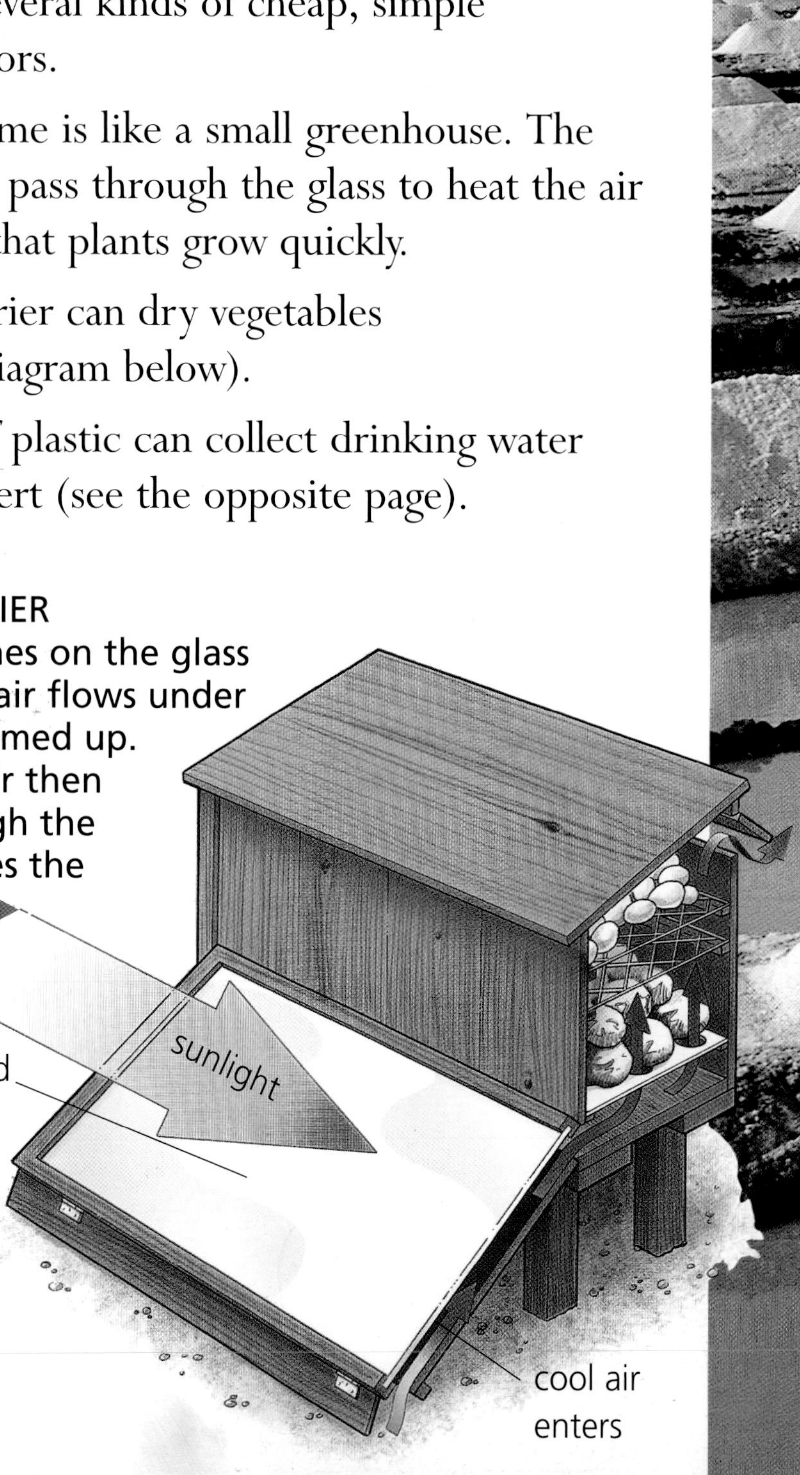

glass cover over blackened sheet

sunlight

cool air enters

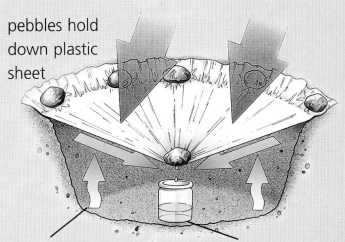

pebbles hold down plastic sheet

dew evaporates

water collected in jar

TO COLLECT WATER IN THE DESERT

● Dig a pit and place a jar in the bottom.

● Stretch a sheet of plastic over the pit in the early morning.

● The moisture condenses on the underside of the sheet and drips into the jar.

◄ Salt water is collected in pools. These pools are called saltpans. The Sun warms the pans. The water evaporates. Salt crystals are left behind.

A typical flat-plate collector

A flat-plate collector is a box covered with glass. The inside of the box is painted black. Water flows through a pipe that snakes through the box. The pipe is also painted black. This is because black absorbs heat well.

How flat-plate collectors work

Well-made flat-plate collectors can heat water to nearly boiling point. They are used to heat water for washing and heating. This is how they work:

- The sunlight heats the box.

- The hot air in the box heats the water in the pipes.

This is a cut-away diagram of a flat-plate collector. As the water pipe snakes through the hot box, the water is heated. ▼

In northern countries flat-plate collectors are on south-facing walls. This is so that they will collect the most light. This flat-plate collector is used to heat water at a hospital in England. ▼

glass cover

inside painted black to absorb the heat

hot water out

cold water in

insulation

This photograph shows two types of flat-plate collector at a research centre. The scientist is checking how well they work. He wants to find out which one is best.

Concentrating collectors

Concentrating collectors collect the Sun's rays over a large area (see the photograph on page 21). Then the sunlight is focused into a small area. This makes very high temperatures (several thousand degrees).

Solar furnaces and heliostats

The type of concentrating collector shown in the photograph on page 21 is a solar furnace. Some of its mirrors move automatically to follow the Sun. These sorts of mirror are called heliostats.

Trough collectors

Trough collectors are more simple. Their mirrors do not move. This means they are less expensive to make but they do not produce such high temperatures.

The curved mirrors concentrate sunlight on to the pipe which is full of water waiting to be heated. ▶

hot water flows out of pipe

sunlight focused on to central pipe

curved mirrors

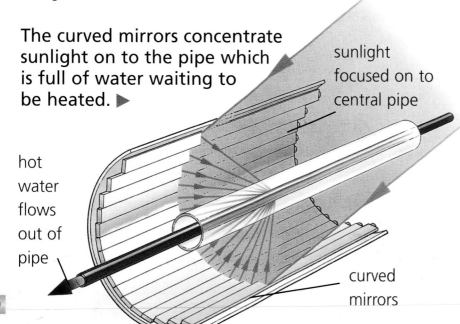

FACTFILE

On the hill facing the huge dish-shaped wall of mirrors are 63 heliostats. Each heliostat is 7.3 metres high. These heliostats automatically follow the Sun. They reflect the Sun's rays on to the dish of mirrors. The dish of mirrors reflects back on to a target only 45 centimetres square. The sunlight is so concentrated that the temperatures can go up to 3,800 degrees Celsius.

Europe's largest solar furnace is at Odeillo in France. This photo shows the huge dish of mirrors (called a parabola). The 63 giant heliostats are out of the photo, facing the parabola.

Facts about Solar One

- Solar One was built in 1982.
- It was the world's biggest solar power station.
- It supplied electricity to 10,000 people.
- It has almost 2,000 mirrors.

How it works

The mirrors reflect sunlight on to a boiler on top of the power tower. The oil in the boiler is heated. It flows to another building. There the hot oil heats water and turns it into steam. The steam turns turbines. This powers a generator to make electricity (see the diagram on page 23).

The mirrors surround a central tower. One name for this type of concentrating collector is a power tower.

HOW THE POWER TOWER WORKS

The mirrors reflect sunlight on to a boiler on top of the power tower.

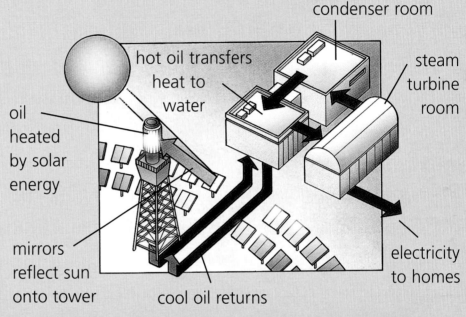

condenser room

hot oil transfers heat to water

steam turbine room

oil heated by solar energy

mirrors reflect sun onto tower

cool oil returns

electricity to homes

At present solar cells only make about 10 per cent of the sunlight falling on them into electricity. But they are being improved. Already they can power some telephones, medical equipment and water pumps.

A LIGHTHOUSE
During the day the solar panels soak up sunlight and make electricity. The electricity is stored in batteries until the lighthouse needs it after dark.

A photovoltaic cell or PV cell

A photovoltaic cell is also called a solar cell. Just one cell the size of a finger nail makes a tiny electric current (see the diagram on the right).

What solar cells can power

- Three or four can power a pocket calculator.
- Several thousand can power a satellite in space.

Good and bad things about solar cells

There is no pollution from solar cells, and sunlight is free. However, the solar cells are expensive and do not produce enough power.

Solar cells are usually made from silicon. The energy from sunlight moves electrons through the cell. This makes an electric current.

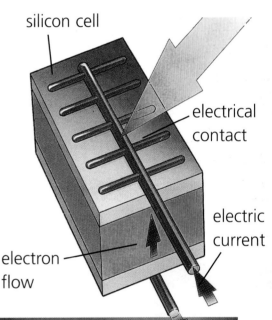

silicon cell

electrical contact

electric current

electron flow

Solar panels make electricity at a solar power station in California, USA. ▶

USING SOLAR ENERGY

Greenhouses

Greenhouses are made of glass. The Sun's rays go through the glass and heat the air inside. The air is then trapped. Gardeners use greenhouses to grow plants that need plenty of warmth.

Other buildings

Office buildings are often built with glass walls, like this one in Texas, USA. ▼

Houses, offices and other buildings can have large windows too. Glass is cheap and it lets in sunlight to help warm the buildings.

An energy-efficient house

Glass lets in the Sun's rays. On a hot day it can be too hot. On a cold day it can be too cold. The house in the diagram shows how sunlight falls on to the big windows. It warms the air inside the house. The warm air rises, and cooler air from the top of the house flows down to the floor. This cool air cools the house. Building lots of energy-saving houses could save millions of tonnes of fossil fuels. Also, carbon dioxide is released when fossil fuels are burnt. If too much carbon dioxide is released the Earth's atmosphere may heat up too much. This is called global warming.

An energy-saving house ▼

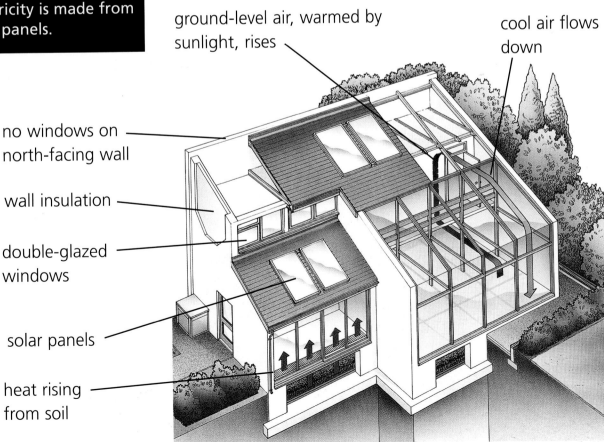

ground-level air, warmed by sunlight, rises

cool air flows down

no windows on north-facing wall

wall insulation

double-glazed windows

solar panels

heat rising from soil

A solar home is a house that uses solar energy efficiently. Solar energy systems can be put into new houses. But old houses can use solar energy too.

An old house in Denmark

The Bille family lived in an old house. It was heated using an oil-burning furnace. Then in 1993 they fitted four solar panels in the roof. These heated all the water in summer and helped to heat water in winter. The family burned 500 litres less oil in a year than they did before.

The solar panels have to be kept clean to get the most energy from the Sun. But only thick snow or very thick clouds stop them working completely. ▼

The white water tank sits next to the old green coal-burning furnace and further to the right is the oil-burning furnace. ▶

cold water to solar collectors

hot water

cold water to oil-fired furnace

hot water from oil-fired furnace

hot water from solar collectors

hot water to taps in house

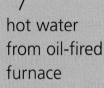

▲ There are two systems that heat the water in the water tank:

- The large coil (heated by solar collectors).

- The small coil (heated by an oil-burning furnace). This cuts in when there is not enough sunlight.

▲ Some solar houses use flat-plate collectors and photovoltaic cells. In this photograph, panels of photovoltaic cells are being made. Silicon for the cells is made from sand and rock.

29

◀ These street lights in Spain are powered by batteries that are charged with electricity from solar panels.

▲ Calculators, torches and fans can be solar powered.

FACTFILE

Bioluminescence is a cold light made by animals such as some beetles, flies, worms, deep-sea fish and also some fungi.

They use it to

● attract a mate

● attract some prey

● scare off attackers

Batteries and solar power

Batteries are the most common portable power supply. But they run down. Solar panels are also portable and last a lot longer.

Portable solar power

Solar power is useful to run radios, telephones and medical equipment in remote areas, where there is no electricity supply. So it is excellent for people such as explorers, scientists and doctors.

Solar-powered lighting

All sorts of lighting equipment uses a mixture of solar power and batteries. The solar panels use solar energy to make electricity. The electricity is stored in batteries. It can then be used when it is dark. This sort of lighting is used all over the world.

▲ In this photo, solar panels are powering the pump, filter and heater to run the swimming pool.

Air pollution

There is lots of air pollution in some cities. Most of this pollution comes from car and truck engines.

Electric cars run on batteries, and do not produce air pollution. However, there are disadvantages to batteries. Theye are heavy, and they have to be charged up with electricity every few kilometres. They also take hours to be charged up.

A solar-powered car would be cleaner and lighter, and it would not have these problems.

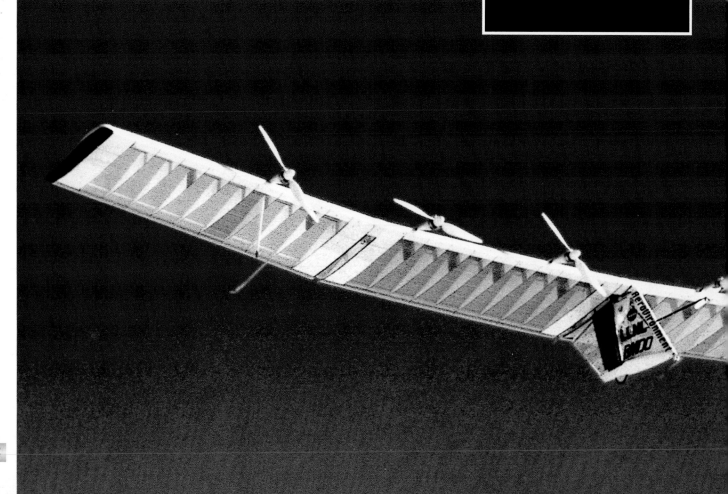

Advantages of solar power

Photovoltaic cells would be excellent to power a car.
- They have no moving parts to wear out.
- They use no fossil fuels.
- They do not need maintenance.

However, photovoltaic cells are not efficient enough to power a family car. They are also very expensive.

Pathfinder is a solar-powered aircraft. It has flown to a height of more than 20,000 metres. ▼

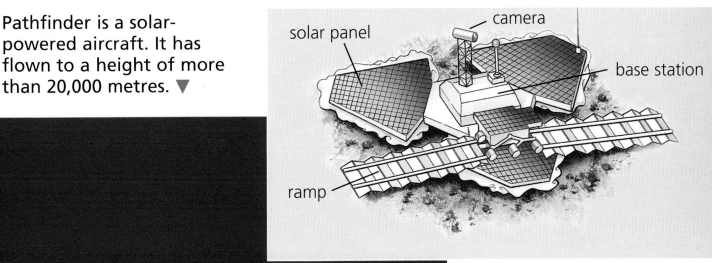

solar panel · camera · base station · ramp

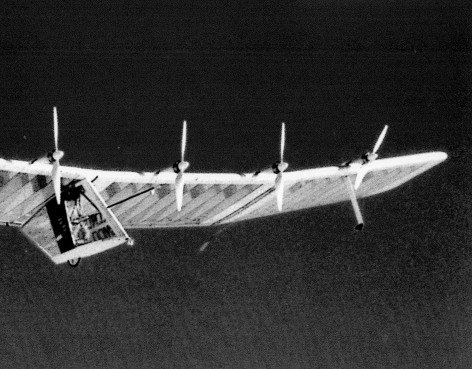

▲ This is a robot called *Sojourner*. It was landed on Mars in 1997. It moved around, took photographs and investigated rocks. Solar panels powered many of its instruments.

This car won the solar car race in 1995. The dark blue top is made up of solar cells. They power its electric motor.

World Solar Challenge car race

Every three years, teams from all over the world meet in Australia to race their solar-powered cars. The race is 3,010 kilometres long. The speed of the cars averages between 67 kph and 90 kph. The purpose of the race is to encourage people to make solar-powered cars.

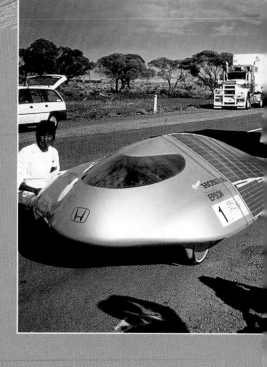

Sunrayce is a solar car race for university teams in the USA. In 1997 the race took 10 days and was nearly 2,000 kilometres long. Sunrayce teams often take their cars to Australia to drive in the World Solar Challenge race.

◀ This car is called *Sunshark*. It raced in the 1996 World Solar Challenge in Australia. The car is controlled with handlebars like those on a bicycle.

Size, shape and silence

Solar cars are small so that they are as light as possible. They are smooth so that they slip through the air. They are silent.

◀ This car was made by Honda. It only weighed 125 kilograms. The solar panels are on the tail.

A Japanese winner

The Japanese car maker Honda recently won the World Solar Challenge in Australia. The car was 5.4 metres long. It was covered with 4,584 solar cells. It had a top speed of 130 kph. However, it cost £700,000 and only two people could sit in it.

The ups and downs of solar power

The amount of solar energy that can be collected goes up and down, as clouds come and go. At night there is no solar energy. So we need to be able to collect solar energy when the Sun is out. Then we need to store it so we can use it later.

Storing solar energy

Solar energy can be stored in batteries. When the Sun goes in, the batteries can power lights, pumps and heating. When the Sun comes out again, the solar energy from the solar panels charges up the batteries.

▲ The solar panels in the background are producing electricity to charge up batteries.

During the day, the Sun warms the Earth. This solar energy controls our weather systems.

Solar ponds

Solar heat can be stored in special ponds. The sides of the pond are thick. This is to stop heat escaping. A typical solar pond is full of salty water. Salty water absorbs heat well.

During the day, sunlight warms the water. The saltiest water sinks to the bottom and holds the heat there. On dull days or at night, the hot water in the pond can be pumped out. It is pumped to a boiler where it is used to make steam. This steam is used to drive a turbine. The turbine powers a generator which makes electricity.

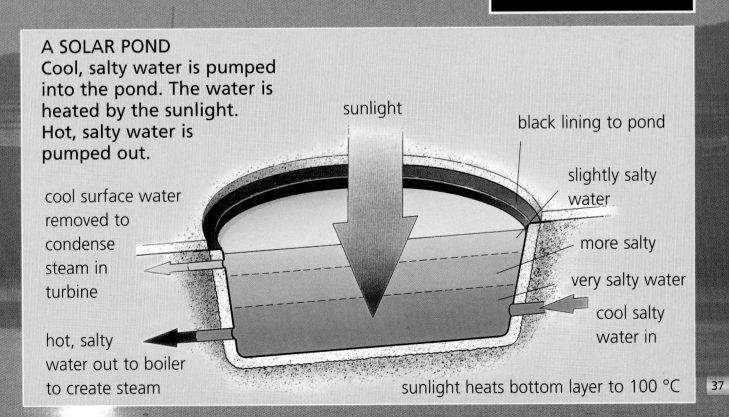

A SOLAR POND
Cool, salty water is pumped into the pond. The water is heated by the sunlight. Hot, salty water is pumped out.

sunlight

black lining to pond

slightly salty water

more salty

very salty water

cool salty water in

cool surface water removed to condense steam in turbine

hot, salty water out to boiler to create steam

sunlight heats bottom layer to 100 °C

A solar-powered telephone booth has been developed. A telephone like this was installed in Ghana in 1997. This sort of telephone can be used by people far away from towns.

A radio station in Mali

Mali is a country in Africa. Parts of Mali are a long way from any big towns. There are no telephones, newspapers or postal services. How do people get information? One answer is radio. But radio stations have to have a reliable electricity supply.

Radio Daande Douentza (RDD) is a solar powered radio station. It started in 1993. For most people it is the only way they get information about health, farming and news. It is partly supported by Oxfam.

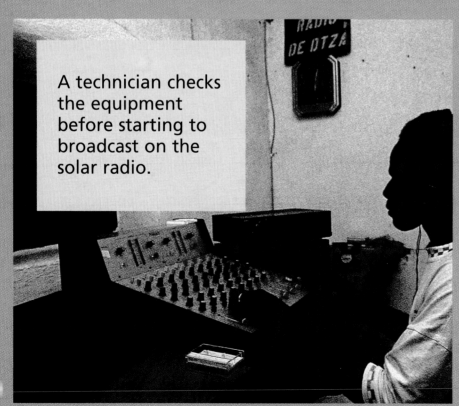

A technician checks the equipment before starting to broadcast on the solar radio.

Radio Daande Douentza uses solar panels to make its electricity. It broadcasts for 56 hours every week.

Success for Radio Daande Douentza

More than 85 per cent of the local people listen to the radio station. In the first six months that it was running, the number of radio sets in the area doubled.

The radio station broadcast information about vaccinating children against disease. The number of children being vaccinated jumped from 30 per cent to 50 per cent of all children in the area.

A radio presenter starts her broadcast.

NEW DEVELOPMENTS

Solar tiles

There are solar panels now that look like ordinary roof tiles. They are about half the cost of other solar panels. This is because they use a different type of silicon, so the roof tiles are cheaper to make.

One third of the roof of an average-size American house could be tiled with solar tiles. Then it would only take six hours of sunshine to run all the electrical goods in the house.

From 1991 to 1993, eight people lived inside this glass building. It was called Biosphere 2. It was in the USA. Heat and light were made by solar energy. The people stayed inside for two years, to test the conditions for a long time. ▼

FACTFILE

Biosphere 2 was the world's biggest closed in living system. There were 3,800 different kinds of plants and animals inside. It was called Biosphere 2 because Biosphere 1 is the Earth itself.

Photosynthesis

All plants have green chlorophyll in their leaves. Chlorophyll is why the leaves are green. The chlorophyll absorbs the sunlight. Then it changes it into energy that the plant can use. This process is called photosynthesis.

If scientists could make an artificial leaf that works like a plant leaf, then solar energy would be much cheaper and more efficient than it is now.

▲ This roof on a building in the USA is covered by special solar tiles. They absorb the sunlight and make electricity.

Deserts

If the cost of making solar cells falls, solar cells could be laid out in hot deserts. They could provide electricity. Where deserts are near the sea, the electricity could be used to turn sea water into drinking water.

But sand blows about in deserts. This could cause two main problems:

- Sand might settle on the mirrors or solar panels. This would block the sunlight.
- Sand might scratch the mirrors or solar panels after a few years. They would not work so well. Glass might have to be used to protect the mirrors and solar panels.

FACTFILE

Solar power stations use so many mirrors and solar panels that they cover large areas of ground even in sunny areas. In less sunny areas they would have to cover even more ground.

This diagram shows how sunlight heats the air in plastic tunnels. The hot air rises up a chimney. As it rises, it turns a turbine that is connected to a generator that makes electricity. ▼

hot air rises and escapes at the top of the chimney

sunlight

turbine driven by rising air

clear plastic tunnel

cool air enters

generator

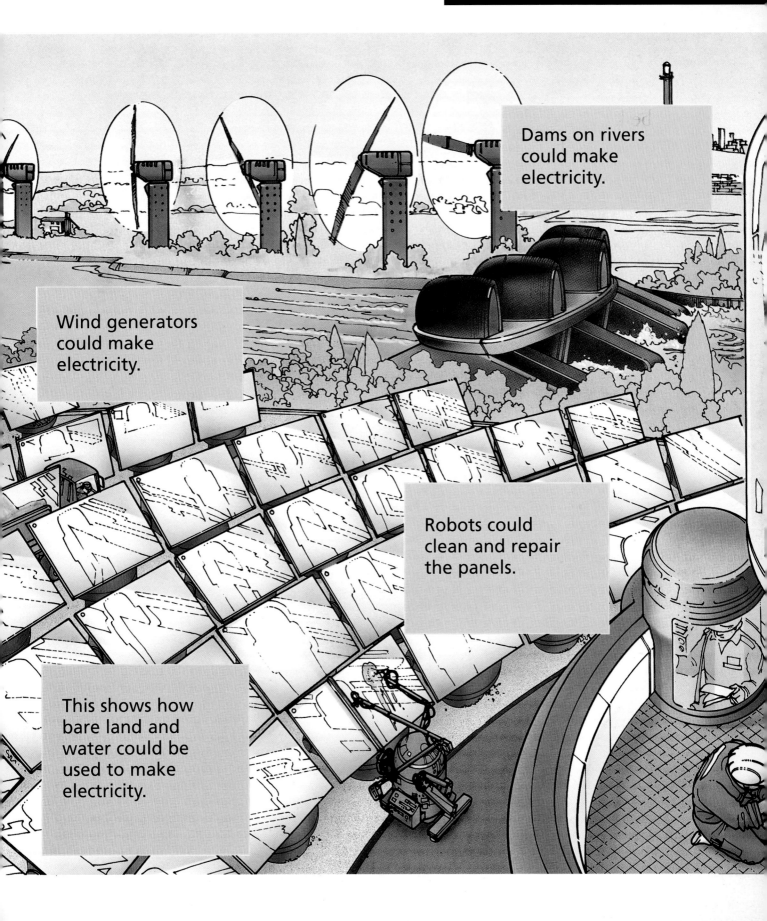

Dams on rivers could make electricity.

Wind generators could make electricity.

Robots could clean and repair the panels.

This shows how bare land and water could be used to make electricity.

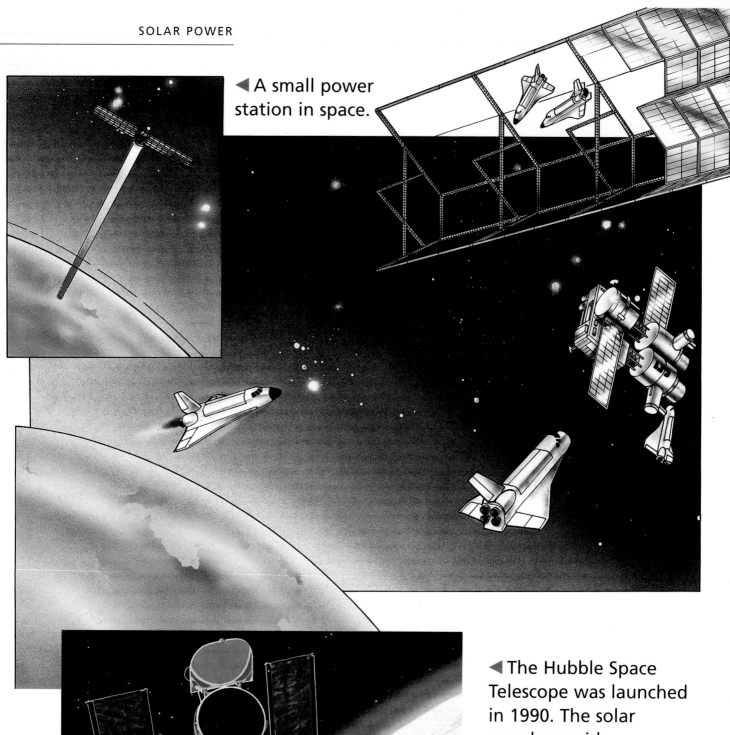

◀ A small power station in space.

◀ The Hubble Space Telescope was launched in 1990. The solar panels provide electricity for its computer and instruments.

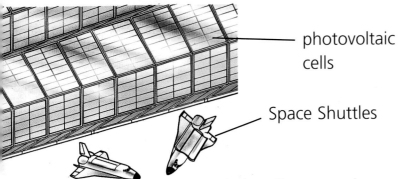

photovoltaic
cells

Space Shuttles

◄ The diagram shows a giant power station of photovoltaic cells in space. The cells would be kept pointing at the Sun by automatic instruments. The power would be beamed down to Earth. Another idea is to have a number of small power stations like the one shown on page 44.

like the one shown on page 44.

Dirty air

The air or atmosphere around the Earth is full of dirt. There is dust from volcanoes and sandstorms. There is pollution from cars, factories and power stations.

A solar collector in space

A solar collector or photovoltaic panel would get much more solar energy if it was outside the Earth's atmosphere. This would mean building a power station in space. It would use solar panels to make electricity. This would be changed into microwaves and sent to Earth. It would be changed back into electricity.

Sunlight sails

Future spaceships might use huge sails, to be blown along by the pressure of the Sun and solar winds.

GLOSSARY

Atmosphere The gases that surround a planet, moon or star.

Atom A very small particle. (A particle is a tiny part of something.)

Aurora The rainbow lights in the sky near the North and South poles.

Carbon dioxide A gas in the air made from plants and animals breathing, and from burning fossil fuels.

Concentrating collector A large mirror or lens that gathers sunlight and focuses it in one spot.

Electromagnetic waves Waves that have both electric and magnetic parts.

Electron A particle with an electric charge.

Element A very simple substance.

Energy Able to do work.

Evaporate To turn from a liquid to a gas.

Flat-plate collector A flat panel that changes sunlight into heat.

Focus The point where light rays come together.

Fossil fuels Fuels formed millions of years ago from the remains of plants and animals.

Helium An element formed inside very hot stars.

Hydrogen The lightest element in the Universe.

Infra-red Electromagnetic waves which transfer heat.

Kilowatt A unit of electrical power (1000 watts).

Magnetic field The area around a magnet where the magnet affects things.

Microwaves Radio waves used for sending signals and for cooking.

Nuclear fission When a heavy nucleus splits apart. This releases a lot of energy.

Nuclear fusion When nuclei hit each other so fast that they join together. This releases a lot of energy.

Nucleus The particle or particles at the centre of an atom.

Photosynthesis A green plant's process of making food using the energy of the Sun.

Photovoltaic cell A device that converts sunlight into electricity.

Recycling Using things again instead of throwing them away.

Silicon An element that is used to make solar cells.

Solar cell Another name for a photovoltaic cell.

Solar collector Anything that receives energy from the Sun and converts it into another form of energy.

Solar energy Energy from the Sun.

Solar furnace A type of concentrating collector.

Solar power station A power station that makes electricity from the Sun's energy.

Solar wind Particles that stream out from the Sun.

Watt A unit of power.

FURTHER INFORMATION

Books to read

Action for the Environment: Energy Supplies by Chris Oxlade and Rufus Bellamy (Franklin Watts, 2006)

Alpha Science: Energy by Sally Morgan (Evans, 1997)

A Closer Look at the Greenhouse Effect by Alex Edmonds (Franklin Watts, 1999)

Cycles in Science: Energy by Peter D. Riley (Heinemann, 1999)

Essential Energy: Energy Alternatives by Robert Snedden (Heinemann, 2006)

Future Tech: Energy by Sally Morgan (Belitha, 1999)

Saving Our World: New Energy Sources by N. Hawkes (Franklin Watts, 2003)

Science Topics: Energy by Chris Oxlade (Heinemann, 1999)

Step-by-Step Science: Energy and Movement by Chris Oxlade (Franklin Watts, 2002)

Sustainable World: Energy by Rob Bowden (Wayland, 2003)

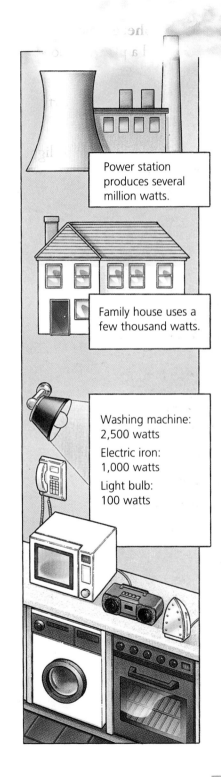

Power station produces several million watts.

Family house uses a few thousand watts.

Washing machine: 2,500 watts

Electric iron: 1,000 watts

Light bulb: 100 watts

ENERGY CONSUMPTION

The use of energy is measured in joules per second, or watts. Different machines use up different amounts of energy. The diagram on the right gives a few examples. ▶

INDEX